FIRE ADMINISTRATION

UNITED STATES

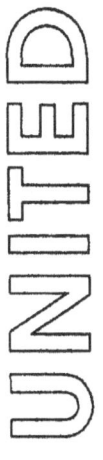

The Crash of Two Subway Trains on the Williamsburg Bridge

June 5, 1995

New York City, New York

TECHNICAL RESCUE INCIDENT REPORT

Federal Emergency Management Agency

This report is part of a series of reports on technical rescue incidents across the United States. Technical rescue has become increasingly recognized as an important element in integrated emergency response. Technical rescue generally includes the following rescue disciplines: confined space rescue, rope rescue, trench/collapse rescue, ice/water rescue, and agricultural and industrial rescue. The intent of these reports is to share information about recent technical rescue incidents with rescuers across the country. The investigation reports, such as this one, provide detailed information about the magnitude and nature of the incidents themselves; how the response to the incidents was carried out and managed; the impact of these incidents on emergency responders and the emergency response systems in the community; and the lessons learned. The U.S. Fire Administration greatly appreciates the cooperation and information it has received from the fire service, county and state officials, and other emergency responders while preparing these reports.

This report was produced under contract EMW-94-C-4436. Any opinions, findings, conclusions, or recommendations expressed in this publication do not necessarily reflect the views of the U.S. Fire Administration or the Federal Emergency Management Agency.

Additional copies of this report can be ordered from the U.S. Fire Administration, 16825 South Seton Avenue, Emmitsburg, MD 21727.

The Crash of Two Subway Trains
on the Williamsburg Bridge
New York City, NY
June 5, 1995

Local Contacts:

Assistant Chief James T, Bullock
 Commanding Officer
FDNY Special Operations Command
750 Main Street
Roosevelt Island, NY 10044

Salvatore Gilardi
 Manager, Fire Safety
NYC Transit Authority
Office of System Safety
370 Jay Street, Room 607
Brooklyn, NY 11201

Deputy Chief Paul Maniscalco
 Commanding Officer
NYC-EMS Special Operations Division
55-30 58th Street
Maspeth, NY 11378

Captain Joseph Murphy
 Commanding Officer
NYPD Office of Emergency Management
One Police Plaza
New York, NY 10038

OVERVIEW

This report details the interagency response of the New York City Emergency Medical Service, Fire Department, and Police Department to the crash of two subway trains on the Williamsburg Bridge on June 5, 1995. Approximately two hundred people were on board the

Manhattan-bound "J" train when it rear-ended a Manhattan-bound "M" train that was stopped on the tracks ahead of the "J" train.

The "J" train penetrated well into the "M" train, crushing the train's cab around the motorman, killing him. Dozens of passengers were injured during the crash, and their medical presentations required using cervical-spine immobilization precautions prior to moving them. The crash occurred on a portion of track that is between the bridge's two roadways and elevated about 15 to 20 feet above them. Patients on backboards had to be lowered to the roadway using Stokes baskets, ladders, and ropes. There is no roadway or solid decking under the tracks to prevent a person or object from falling 133 feet to the East River; a fall between the track ties would have been fatal.

Key findings of this investigation are:

- A unified command utilizing the Incident Command System was implemented early into the incident by the participating agencies. This facilitated effective control of the incident and contributed to a rapid resolution of the emergency without emergency responders sustaining any serious injuries.

- Technical rescue does not always involve complicated equipment, systems, and procedures. Safe and efficient patient evacuation was performed using ropes, ladders, and Stokes baskets in a simple, yet effective, patient lowering system.

- Interagency communication was effective and demonstrated the benefits of the frequent disaster exercises in which all of the key emergency response agencies have participated.

- Prior interagency planning and exercises allowed the incident command personnel for the different agencies to develop personal relationships with their counterparts. This promoted interagency coordination and cooperation and made each person's job easier.

- Construction contractors who were working on the bridge provided resources and expertise that facilitated the accomplishment of patient and passenger evacuation.

- Populated areas beneath the bridge were evacuated to prevent falling debris from injuring any civilians or responders below the crash site.

Technical Rescue Incident Investigation Project Explained

The Technical Rescue Incident Investigation Project is an effort of the U.S. Fire Administration to document case studies of certain technical rescue incidents. The project seeks to produce documentation in a "lessons learned" format in order to provide local emergency responders, trainers, federal and state agencies, and other interested groups enhanced knowledge about technical rescue response and safety.

Applicability to Williamsburg Bridge Subway Crash

The safe and efficient rescue of over 200 passengers from the two subway trains high above the East River and Brooklyn shoreline demanded the highest level of interagency cooperation and technical skill from the three major responding agencies. The successful outcome of this incident is directly related to the use of the Incident Command System, meaningful and frequent interagency drills, ingenuity in the application of rescue skills, and an overriding concern for patient and rescuer safety. Accordingly, the management of this incident serves as an excellent case study for the fire service to understand the value of close control of technical rescues such as this.

Investigation Methodology

The research for this report was conducted through face-to-face interviews with several individuals who had incident command and operational responsibilities during the response to the Williamsburg Bridge subway crash. In addition to the interviews, documentation was provided by the New York City Emergency Medical Service (NYC-EMS), New York City Fire Department (FDNY), New York City Police Department (NYPD), and New York City Transit Authority (NYCTA). Finally, a site visit was made to view the scene of the crash.

Acknowledgements

This report could not have been written without the invaluable assistance of the New York City Emergency Medical Service, the New York City Fire Department, the New York City Police Department, the New York City Transit Authority, and the New York City Department of Transportation Division of Bridges and Roadways.

◆ ◆ ◆

I. Introduction

New York City is the largest city in the United States, with a population of 7,322,564 (as of the 1990 Census). Emergency services are provided by three separate agencies, the New York City Emergency Medical Service (NYC-EMS), the New York City Fire Department (FDNY), and the New York City Police Department (NYPD).[1] Vehicle extrication is generally provided by the NYPD's Emergency Service Unit (ESU); however, the FDNY also maintains a technical rescue capability.[2] The determination of which agency's technical rescue services will be utilized at a given incident depends on the nature of the situation and is often decided on a case-by-case basis.

The collision of two subway cars on a bridge, approximately 130 feet above New York City's East River (and partly over the Brooklyn shore), produced numerous patients who needed to be rapidly and safely backboarded and lowered from the railbed to an adjacent roadway before they could be transported by ambulance to the hospital. The risks to rescuer safety included falls from heights, contact with electric current, and cuts from the wreckage.

Rescuers from all of the agencies demonstrated excellent teamwork, ingenuity, and technical skills in clearing the trains of injured passengers. All passengers were treated, packaged, evacuated and transported within about three-and-one-half hours. Ambulance access to the congested bridge was assured through careful and early consideration of this necessity by the Incident Command Team. No single hospital received a disproportionate share of the patients (see Table 1 for a breakdown of patient destinations). Responder injuries were limited to two minor injures to police officers.

---

[1]Throughout this document, references to the services are given in alphabetical order.

[2]The NYC Transit Police formerly had their own ESU which specialized in extrication related to subway train mishaps; however, this unit was merged into the NYPD's ESU when the Transit Police became part of the NYPD in March 1995.

Table I Number of Patients Transported by Destination

Destination	# of Patients
Queens	
Queens Hospital Center	8
Elmhurst Hospital Center	5
St. John's Hospital	2
Brooklyn	
King's County Hospital	8
Long Island College Hospital	6
Methodist Hospital	2
Wyckoff Heights Hospital	4
Woodhull Hospital	10
Brooklyn Hospital	6
Manhattan	
Bellevue Hospital	3
St. Vincent's Hospital	1
NYC Morgue	1

II.   Description of the Crash[3]

At approximately 0612 hours, Monday, June 5, 1995, the Manhattan-bound 0531-J train[4] collided with the Manhattan-bound 0548-M train, which was stopped for a red signal on the J2

3At the time of the publication of this report, the accident was under investigation by the National Transportation Safety Board, and hence no official cause had yet been determined.

4NYC subway trains are designated by the time which they depart their station of origin. The 0531-J departed the Jamaica Center terminal at 0531 hours. Likewise, the 0548-M departed the Metropolitan Avenue terminal at 0548 hours.

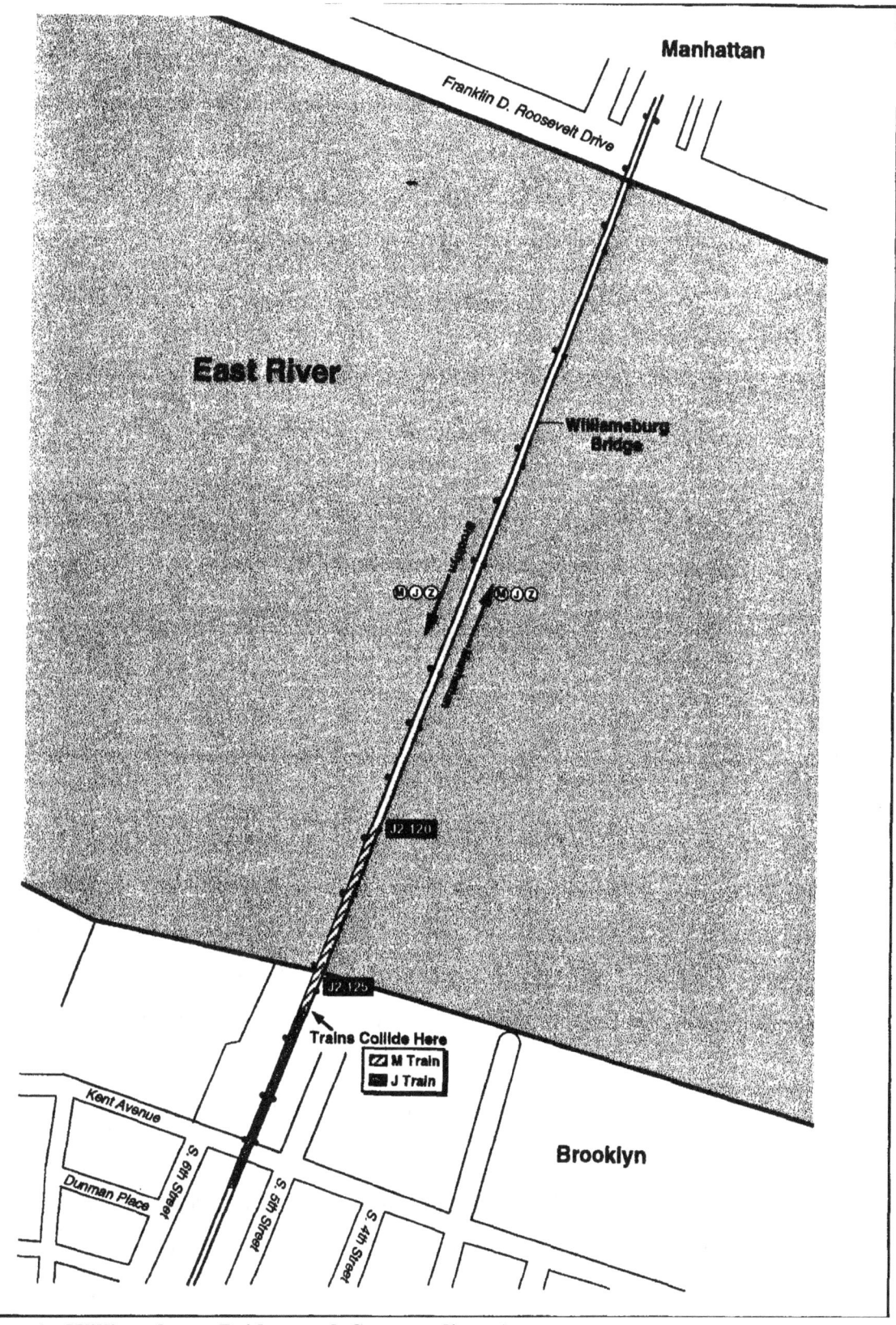

Figure 1. Williamsburg Bridge and Surrounding Area

track, about 16 feet south of signal 125 at the east end of the Williamsburg Bridge (see Figure 1 for a map detailing the crash location). The collision occurred between the Brooklyn east tower and the Brooklyn anchorage (structural elements of the bridge). The M tram was partly over land and partly over water, and the J train was over land. National Atmospheric and Oceanographic Administration maps indicate that there is a clearance of approximately 133 feet from the bottom of the bridge to the mean high water mark below the where the M train had stopped (see Figure 2 for an elevation of the bridge).

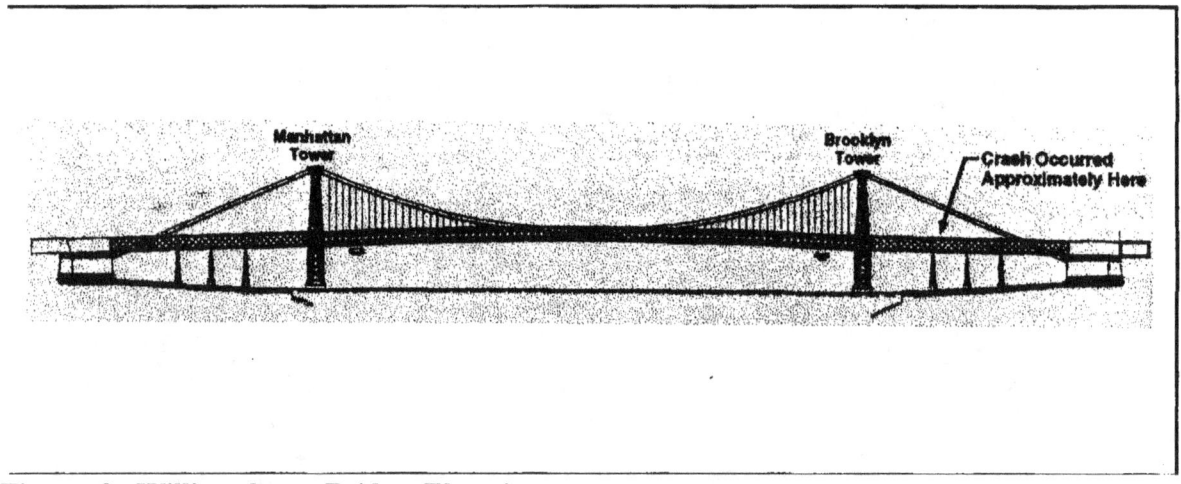

Figure 2. Williamsburg Bridge Elevation

The subway tracks occupy the center area of the bridge flanked by the east- and west-bound roadways. To the north and the south of the subway tracks, and about six feet above them, two pedestrian walkways with wood plank decks each cover one of the vehicle roadways that are 15 to 20 feet lower than the tracks (see Figure 3 for a cross-section of the bridge structure and Figure 4 for a plan showing the relationship between the roadways and the tracks).

At the time of the crash, the area of the bridge above and adjacent to the trackbed was undergoing structural repairs. Due to the construction, the south pedestrian walkway was closed and its planking had been removed; the north walkway was open for pedestrian traffic. The bridge repairs did not interfere with subway operations, but as shall be described later in this report, the presence of bridge workers and construction materials played a helpful role in the rescue operations.

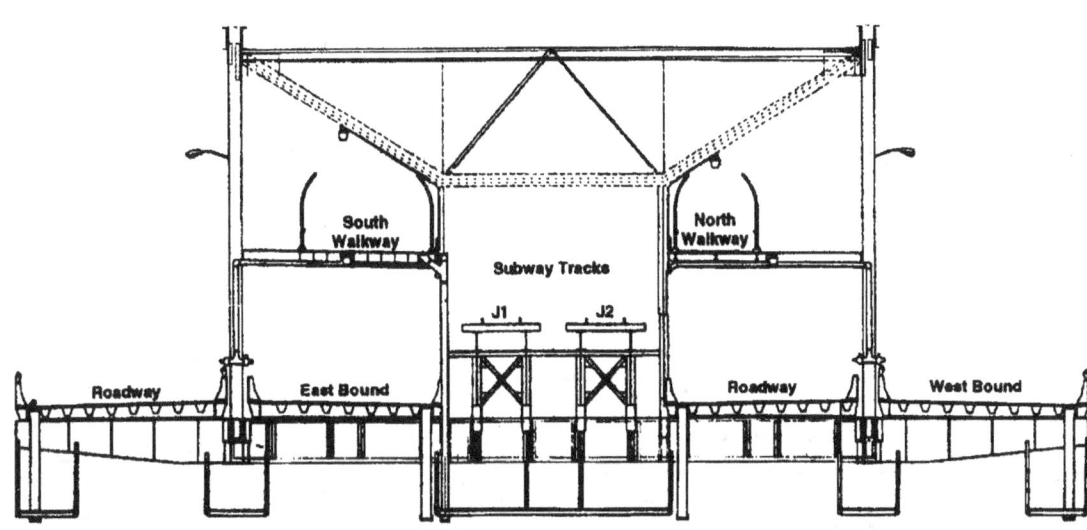

Figure 3. Williamsburg Bridge Cross-Section

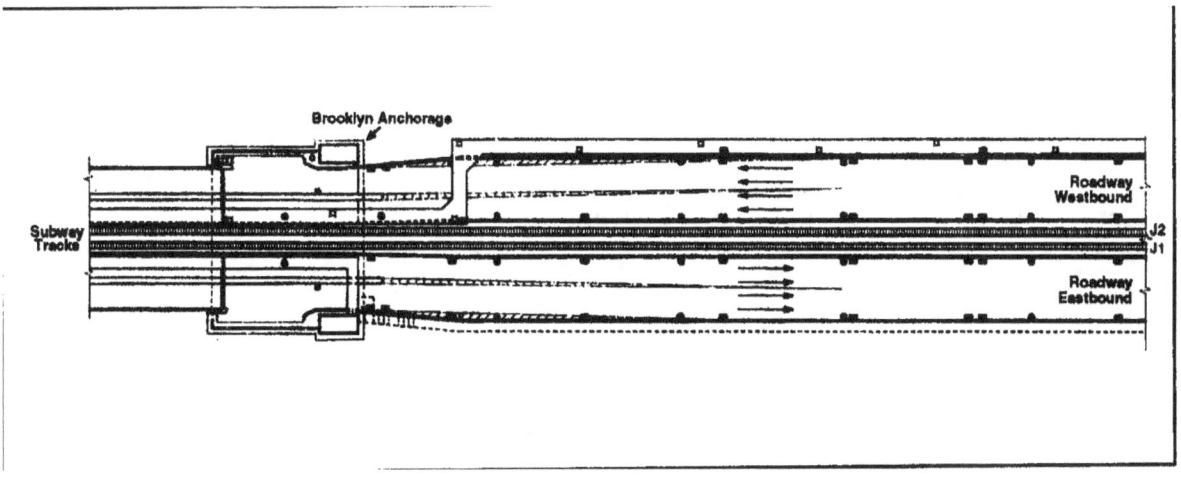

Figure 4. Williamsburg Bridge Plan of Roadways and Subway Tracks

The front of the J train struck and penetrated the stopped M train, resulting in a combined crush of about 14 feet. No derailment or fire occurred as a result of the crash, but there was extensive damage to the lead car of the J train (#4461) and the last car of the M train (#4664). The motorman of the J train was killed instantly and entrapped in the crushed wreckage of his motorman's cab. Sixty-three passengers and the J train conductor were injured in the collision (two police officers received minor injuries during the ensuing rescue).

9

The J train consisted of eight R40-type subway cars. The M train consisted of eight R42-type subway cars. Each of these cars has a capacity of 44 seated passengers and approximately 225 standing passengers (see Figure 5 for a diagram of the R40- and R42-type subway cars).

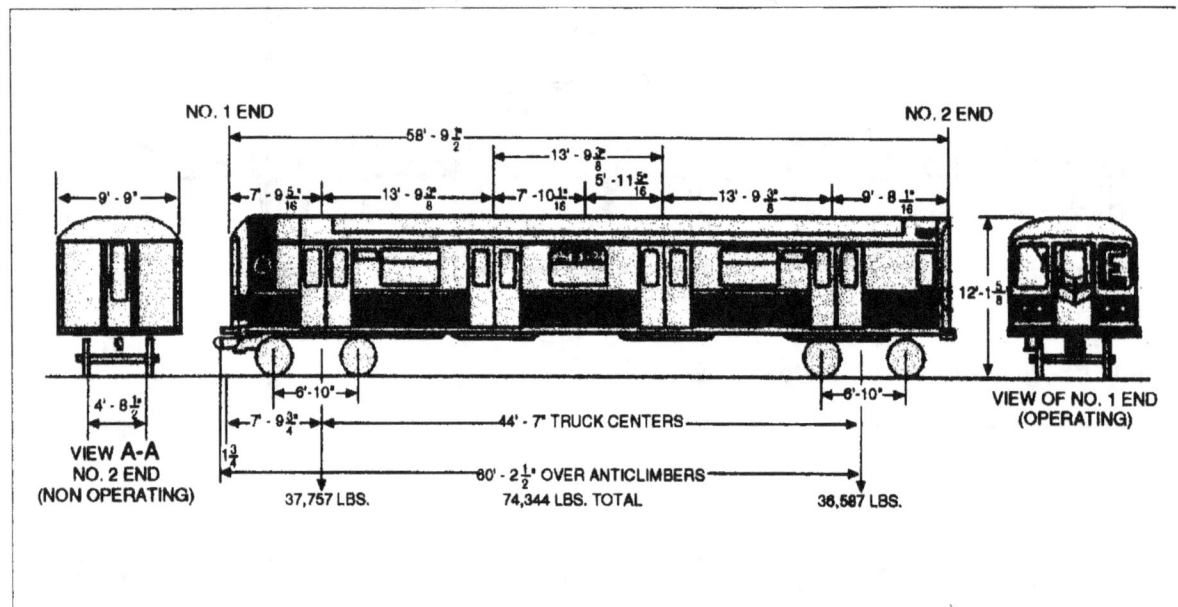

Figure 5. R-40 and R-42 Subway Car Layout

III. Establishment of Command

A unified command was established in accordance with NYCTA's Policy Instruction 8.1.2, "Procedures for Response to Rapid Transit Emergencies." This document details the actions that will be taken in response to a wide variety of specified emergencies.[5] The FDNY, NYC-EMS, and NYPD have all reviewed and agreed to the procedures, and all three are signatories to the document.

---

[5]The policy instruction does not, however, give any guidance on how to handle emergencies on bridges. Subsequent to this incident, it was decided that this condition needed to be corrected, and efforts are underway to prepare an additional section for the manual on proper response to bridge emergencies.

By all accounts, much of the success of the operation can be related to the early establishment of command and institution of an Incident Command System organizational structure. This allowed commanders from the three agencies to make well-informed decisions with input from the other agencies present. This was critical in establishing priorities, such as preserving ambulance access to the bridge roadway. Once a command decision was made, all of the affected agencies knew about it and could provide accurate and timely information to their personnel.

One of the first sectors established by the NYC-EMS commander was triage. Following the standard operating procedure for multiple casualty incidents (see Appendix A), the START triage system was used to identify the order in which patients were to be treated and transported. EMS personnel reported that familiarity with the Simple Triage and Rapid Treatment (START) triage system, because of its uniform use on major incidents, allowed them to implement triage without delay.

IV. Rescue Techniques

The term "technical rescue" is often associated with operations that require complex equipment, techniques, and tactics. In this case, however, only basic ropes, pulleys, carabiners, and know-how were used to effect the rescues. The "technical" rescue involved hand-lowering patients in Stokes baskets down from the trackbed to the roadway below. This provides an excellent example of how simple techniques can be combined and applied to a situation in order to produce the wanted results.

Backboarded patients were placed into Stokes baskets. The loaded baskets were then lowered against 16-foot fire department ladders which had been placed so as to provide a stable surface to guide the basket down (see Figure 6 for a photograph of the placement of the ladder and the Stokes basket). No mechanical advantage hauling system was used to get the baskets up

Figure 6. Patient being lowered from the trackbed to the roadway

or down. It was thought that the increase in mechanical advantage which would have been afforded by a Z-drag or piggyback hauling system was not needed and would have required an unnecessary amount of time and effort to set up, especially given the relatively short distance that patients needed to be lowered. Instead, firefighters used a simple belay system and their own body weight to counterbalance the loaded Stokes baskets.

This system was fast and easy to set up and did not require advanced technical expertise or fancy equipment. The materials which were used are commonly found on ladder and rescue companies throughout the United States. This successful rescue operation underscores the fact that there is no advantage to be gained by using a complicated system when a simple approach will do the job just as well and just as fast.

The only extrication work that was necessary was the recovery of the motorman's body. The impact had so severely crushed the front cab of the J train that the initial responders quickly realized that the crash had been "non-survivable" for the motorman. Responders had to work with mechanical spreaders and cutters for over twenty minutes to remove this victim.

V.   Safety Considerations

The location of the incident, high above the East River and the Brooklyn shore, created two special safety concerns. First, unlike a conventional at-grade trackbed, there was little structure beneath the elevated trackbed to prevent a fall all the way to the river or ground below. People walking in the area had to be extremely cautious to avoid falling between the ties or through gaps in the structure. The construction crews working on the bridge in the area of the crash assisted by providing plywood which was used to cover areas where passengers and responders needed to traverse. The NYPD assigned a launch to stand by in the river below the bridge in case it became necessary to rescue someone who had fallen into the water.

Because they were unsure of the extent of damage to the undercarriage of the subway cars, the incident commanders ordered the evacuation of all land areas under the crash site to guard against injuries from falling debris. Although no debris ultimately fell, this is an excellent example of forward thinking and contingency planning.

Additional safety actions included FDNY's stretching a hoseline to the trackbed. Although there was no initial report of fire, prudent incident management dictated that firefighters be prepared for the risk of a fire during the removal of passengers and the extrication of the train operator. Ultimately, no fire suppression efforts were necessary.

The NYPD aviation unit contacted the Federal Aviation Administration to have the airspace surrounding the incident restricted, in order to prevent aerial "rubbernecking" (a large number of helicopters regularly operate in New York City). The NYPD harbor launch which was standing by below the bridge also kept other watercraft away from the danger zone. This rescue was unusual because it was necessary to control access to the scene from more than one dimension.

Both the FDNY and NYC-EMS had safety officers assigned to the incident and functioning within the Incident Command System. The two on-duty Safety Battalion Chiefs were dispatched to the incident by the FDNY (although one later left to handle a serious firefighter injury at another incident). The FDNY Safety Battalion Chiefs receive special training as incident safety officers and respond to all major incidents with the authority to halt and correct any unsafe actions. NYC-EMS assigned two command-level officers to handle safety, one at the trackbed and one on the west-bound roadway.

VI. Preparedness

The NYCTA conducts four functional exercises each year, each involving the three major emergency agencies. The previous quarterly drill was conducted on March 28, 1995, and NYC-

EMS had conducted a Mass Casualty Incident drill only six days earlier, on May 31, 1995. The NYC-EMS exercise simulated simultaneous NYC-EMS response to an overturned school bus (with 15 patients) and a multiple-alarm fire (with 15 patients). All of the responders indicated that regular drills had provided a stable framework for executing large-scale responses and offered a valuable opportunity to get to know and work with their counterparts in the other agencies. There was consensus that these prior training events played a major role in the smooth execution of the Williamsburg Bridge rescue.

VII. Discussion

The number of casualties from this incident was low because the crash occurred early in the morning. An hour later, the two trains would have been packed to capacity with rush hour commuters on their way to work in Manhattan. There were only about 200 passengers on the two trains at 0612 hours. At capacity, there could have been more than 4,000 passengers on board. It is fortunate that the crash occurred when it did.

The efforts of the responders to this difficult situation deserve commendation. All the passengers who presented any complaint of neck or back pain were evacuated from a dangerous location utilizing proper spinal precautions. Only two responders received even minor injuries. The incident was concluded and subway and vehicular traffic was restored to a critical transportation artery by mid-afternoon.

Patient Accountability and Transportation

Given prior experience with incidents involving large numbers of patients and public transportation, the responders knew they had to deal with two patient accountability concerns. Ensuring that no single hospital bore a disproportionate load of patients was a primary concern (see Table 1 for a breakdown of patient destinations). This would have been even more important had there been more patients requiring high-level trauma services. As it turned out, most of those transported by NYC-EMS ultimately were treated and released from the hospital.

It was also important for legal reasons to document who was taken off the trains, including whether the person was complaining of any injuries at the time of evacuation. This is a critical task, not only for tracking which patient went to which hospital (for the purposes of reuniting family members and for follow-up investigations), but also for establishing who was actually on the train (or bus).[6]

VIII. Lessons Learned

The following were cited by responders as critical lessons learned to be shared with other emergency responders:

- Familiarity with the other agencies and their incident command staffs is a significant benefit when handling an incident of unusual size or complexity. The participants should understand each agency's mission, priorities, capabilities, limitations, and standard operating procedures. This can only be accomplished through pre-incident planning and regular inter-agency exercises. Familiarity with the individuals as well as the organizations is very helpful; as one responder indicated, "A personal relationship equals instant recognition."

---

[6] Public transportation systems have had many false insurance claims from people who were later shown not to have been on the conveyance at the time of the mishap.

16

- Consider bridges as target hazards for pre-incident planning. This includes determining the best access routes to and from the bridge, learning about catwalks and pathways which are both above and beneath the bridge, identifying electrical or utility lines on the bridge, locating water supplies for fire suppression units, understanding the traffic patterns and periods of congestion on the bridge, and knowing the important structural features of the bridge.

- Keep unnecessary vehicles away from the scene. This is especially important when access to the scene is tight, as in this incident. Units that are not requested should not respond, especially when access to the scene is a problem.

- Think in advance of how patients will be transported. Dispatch ambulances to a staging area at an advantageous location and provide routes to and from the scene. Do not congest the loading area with ambulances waiting for patients. Park vehicles that will not be needed out of the way.

- Keep the news media informed and controlled. Providing accurate and timely information keeps the media from hampering rescue operations as they seek information. Provide "photo opportunities" for still and video photographers. The media can be of assistance in getting important messages to the public (e.g., areas to stay away from, precautionary actions to take, etc.). At the Williamsburg Bridge incident, the press was handled by the NYPD Deputy Commissioner for Public Information and the Mayor's Press Office. This ensured that the press didn't hamper agency commanders or rescue operations.

- Utilize available resources found on the scene when necessary. In this case, the bridge workers (and their expertise) as well as some of the construction materials on site were used effectively by emergency responders to improve the safety of people walking on the tracks and surrounding areas.

- Institute the Incident Command System as early as feasible. In all but the rarest of cases, this will mean that the first-arriving unit will establish or pass command.

- Keep in mind that rail incidents can pose hazardous materials problems, even if the trains involved are simply passenger trains. There are a variety of hazardous materials (such as oil, fuels, etc.) that can be released, even in subway crashes. The Department of Environmental Protection (or its equivalent) can provide valuable control and clean-up advice, either at the scene or via phone.

- Be sure to protect rescuers and victims from the elements. Had it been raining, snowing, very hot, or very cold, considerations would have been made for shelter. In this case, the weather conditions were ideal, and such considerations were unnecessary.

◆ ◆ ◆

Appendix A:
NYC*EMS Multiple Casualty Incident Protocol Information

$Q_{A/I}$

The Cutting Edge

A Quality Assessment/Improvement Publication
NYC-EMS-Division of HHC
vol 4 issue 5

May 1995

MULTIPLE CASUALTY INCIDENT

Due to the increased acts of terrorism and the need for preparedness, this issue of The *Cutting Edge* will address the Emergency Medical Action Plan (E.M.A.P.) and Simple Triage And Rapid treatment (S.T.A.R.T.). EMAP was developed in order to provide for efficient use and distribution of NYC*EMS resources, as well as those of other agencies, which can support prehospital emergency medical care roles. This allows for optimal utilization of resources without impacting on the citywide response. EMAP is divided into seven segments, *Initial Notification, Implementation, Operations, Escalation, Stabilization, De-escalation,* and *Termination.*

The INITIAL NOTIFICATION to Communications, begins when the Call Receiving Operator (CR0) obtains the initial information. The CR0 assigns a call type and relays the information to the Tour Supervisor, who in turn notifies the Tour Commander. The assignment will remain with its original call type until the pre-dispatch information, obtained (from SOD, FIRE, NYPD, etc.) by the Tour Commander, warrants a change in designation to an MCI code. Another manner by which a call type can be changed is when the unit arriving on the scene confirms that the assignment meets any of the following criteria:

1) six (6) or more patients, or
2) five (6) or fewer patients with unusual circumstances, or
3) when the scope of any incident exceeds the capabilities of a two (2) ambulance response.

During the IMPLEMENTATION phase, the EMS Tour Commander (TC) activates EMAP based on the pre-dispatch information. The TC is guided by the Initial Unit Response Matrix (located in the Operating Guide: EMAP C-I). The TC considers the source and accuracy of the information obtained from other agencies, as well as reports from EMS units arriving on the scene, and determines the appropriate radio code signal which most closely descr bes the incident. The TC also insures that appropriate personnel throughout the chain of commend are notified, end that a list of medical facilities closest to the incident is formulated with a bed availability status for each site. The minimum response which can be sent to an MCI is one (1) ambulance and one (1) EMS Officer. The Implementation phase of EMAP is concluded with the arrival of the first EMS unit.

The OPERATIONS phase begins with the arrival of the first EMS unit. The first arriving unit advises the nature of the incident, and establishes a Command Post and Staging Area.

The Vehicle Operator remains with the vehicle and assumes the responsibilities of Communications Officer, until relieved by a Supervisor. The Vehicle Operator relays ail pertinent information to the Dispatcher *as* soon as it is obtained, but no later than five (5) minutes after arrival. if the incident involves explosive materiels or chemicals, *no verbal or modat radio transmissions* closer than 300 feet from the scene shall be made unless clearance to do so has been granted by competent authority. A telephone communication shall be established with the Citywide Dispatcher.

The Technician of the first arriving Ambulance assumes the respons bilities of both the Triage and Treatment Sector Officer and initiates appropriate triage/treatment.

Upon arrival, the most senior paremedic of the first arriving ALS unit at the incident assumes the responsibilities of the Triage and Treatment Sector Officer (Operating Guide 106-2).

The first arriving ambulance shall not transport patients until additional resources have arrived or no patients remain at the scene. Members assigned to the second arriving ambulance shall respond directly to the Staging Area and will make radio contact with the Communications Officer Ipreferably on the command radio frequency). They shall assist the Technician of the first arriving ambulance with Primary Assessment Surveys using Simple Triage and Rapid Treatment (START), or other patient care-related functions. The Operations phase continues until the Incident Commander, (the highest ranking EMS Officer), secures the scene.

START was developed in order to rapidly triage multiple patients at the scene of an incident in order to maximize efficient utilization of resources. START is based on the evaluation of respiration, circulation, and mental status. This evaluation places a patient into a color category, which ultimately determines the priority in which a patient will be removed to a hospital,

* In assessing respirations, if the patient is not breathing after one (1) attempt to reposition the airway, the patient is categorized with a BLACK tag and is considered non-salvageable.

* If, after one (1) attempt to reposition the airway, the patient begins breathing spontaneously, the patient is categorized with a RED tag and will be transported as soon resources are available.

* If the patient is breathing at a rate greater than (>)30 breaths per minute, the patient will be categorized with a RED tag.

* If the patient is breathing at a rate less than (<) 30 breaths per minute, the patient's capillary refill will be assessed. If the patient has a capillary refill delay greater than (>) 2 seconds, then the patient is categorized with a RED tag.
  If the patient has a capillary refill less than (<) 2 seconds, then the mental status is to be assessed. If the patient fails to follow simple commands, then the patient is categorized with a RED tag.

* If the patient follows simple commands, then the patient can be categorized with either a YELLOW tag, if the patient's injuries prevents the patient from walking, or a GREEN tag if the patient is ambulatory.

It is important to remember that the triage process is *dynamic* and should continue throughout patient contact. After the initial START assessment and as resources allow. more detailed examinations may influence triage categorization.

The ESCALATION phase includes all elements of additional resource responses, including major citywide re deployment, and Mutual Aid Responses. The authority to activate the Escalation phase rests entirely with the CO, based on the number of patients and amount of resources operating at the scene. The TC can, prior to the arrival of the CO, escalate the response based on the information obtained from other agencies (FDNY, NYPD, etc.). The Escalation phase runs concurrently with the Operations phase and terminates upon determination by the CO that additional resources are no longer necessary.

In the STABILIZATION phase, an adequate amount of resources are available and deployed at the scene to address the prehospital care needs.

The DE-ESCALATION phase is a deliberate and controlled reduction in personnel or equipment resources. When the EMS resources at an incident exceeds requirements, the De-escalation phase may be considered. This reduction is contingent upon the amount of resources required at the scene for continued EMS operations.

The TERMINATION phase of the incident occurs when the CO determines that there are no patients remaining at the scene and that there is no significant potential for further injury. The EMS CO shall be the last unit to leave the scene, at which time the scene will be considered secured.

It is important to have a good understanding of the function of EMAP in order to maintain safety for all members of the Service operating on the scene of an MCI. A good working knowledge of START will ultimately provide all patients on the scene with the most appropriate and timely prehospital medical care.

For your convenience a wallet card is being provided for quick reference.

This issue of The Cutting Edge is dedicated to all the survivors, patient care providers, and rescuers in Oklahoma City. The Men, Women, and Children who died, as well as their families, will be forever etched in our memory.

It is with pride that we thank our own EMS USAR TEAM who responded to Oklahoma City.

Capt. Charles Wells EMT-P, Lt. Carl Tramontano EMT-P, Lt. James Booth EMT-P, Paramedic Raymond Bonner, Dr. Dario Gonzalez, and Dr. Mark Kindeschuh.

For further information regarding this topic, or any other topic that appeared in The Cutting Edge, please contact the
Office of Quality Assessment & Improvement/Office (710) 4 1 6 - 7 3 3 0
New York City "Emergency Medical Service

♦ ♦ ♦

Appendix B:

Glossary

ESU- Emergency Service Unit (NYPD rescue squad)

FDNY -- Fire Department, City of New York

NTSB -- National Transportation Safety Board

NYC-EMS -- New York City Emergency Medical Service

NYCOEM-- New York City Office of Emergency Management (NYPD)

NYCTA -- New York City Transit Authority

NYPD -- New York City Police Department

Piggyback hauling system -- A system of carabiners, ropes, pulleys, and ascending devices which is configured to provide additional mechanical advantage (beyond that which a Z-drag can provide) for raising or lowering.

RTO -- Rapid Transit Operations (Command Center)

Z-drag hauling system -- A system of carabiners, ropes, pulleys, and ascending devices which is configured to provide a 2:1 or 3:1 mechanical advantage for raising or lowering.

◆ ◆ ◆

Appendix C:

Incident Chronology

0612   0531-J train strikes 0548-M train which is stopped at signal J2.120, waiting for a red light to clear. Operator of M train radios initial report to RTO.

0615   NYCTA RTO removes power to the affected section of tracks (Havemeyer to Chambers Streets).

0617   FDNY receives phone alarm of collision on Williamsburg Bridge, Box 8108. NYPD receives notification of collision. NYPD ESU Truck 8 responds.

0618   FDNY dispatches 4 engines, 2 ladders, 1 Battalion Chief, 1 Deputy Chief, and 1 rescue truck to Box 8108.

0620   NYC-EMS dispatches first two assignments -- a trauma assignment to the Brooklyn side of the Williamsburg Bridge and a standby assignment to the Manhattan side. The EMS Tour Commander upgrades the incident to an EMS Signal 10-31, Multiple Casualty Incident (MCI) (see Appendix A for a description of the NYC-EMS MCI policy).

0621   First NYPD units on scene.

0622   FDNY Engine 15 and Engine 55 arrive on scene. NYPD ESU 8 arrives on scene.

0624   FDNY Ladder 104 and Battalion Chief 4 arrive on scene. FDNY field command post established. NYPD requests 10 ambulances, 5 to each side of bridge.

0625   NYC-EMS llB1 arrives on scene.

0626   NYC-EMS established triage area.

0628   NYPD confirms multiple serious injuries and one DOA.

0630   NYPD shuts down traffic on Williamsburg Bridge.

0633   NYC-EMS "Williamsburg Command" established. Power confirmed off.

0710   Emergency Operations Center at NYC-EMS HQ established.

0716   RTO notified that operator of J train is deceased.

0728   First patient transported.

0752   Passengers aboard M train start to be evacuated.

0838   M train evacuation is completed. NYCTA buses are on scene.

0911   Uninjured passengers of J train start to be evacuated.

0930   J train evacuation is completed. Passengers are aboard buses.

0945   De-escalation of NYC-EMS resources begins.

1001   Last patient transported. Non-injured removed by NYCTA buses with NYC-EMS personnel on board monitoring passengers. NYC-EMS standby operations continue to support emergency responders still operating at the scene.

1100   EMS Signal 10-31 secured.

1445   NTSB releases all trains.

1300   Power and normal operations resumed.

◆ ◆ ◆

*U.S. GOVERNMENT PRINTING OFFICE: 1996-719-595/82752

www.ingramcontent.com/pod-product-compliance
Lightning Source LLC
Chambersburg PA
CBHW081307170526
45165CB00010B/3287